Abacus Arithmetic

1 9 5 8

ABACUS ARITHMETIC

How to Perform Some Calculations on the Chinese Abacus

By

Welton J. Crook

PACIFIC BOOKS
Palo Alto, California

LIBRARY OF CONGRESS CATALOG CARD NUMBER 58–7709

PRINTED AND BOUND IN THE UNITED STATES OF AMERICA

Published by PACIFIC BOOKS, PALO ALTO, CALIFORNIA

Preface

While on visits to the Orient, my wife urged me to purchase an abacus and to learn to use it. Somewhere I read, or was told, that a non-Oriental would have considerable difficulty in learning how to operate the instrument unless he started as a child and grew up with it. However, having discovered a store in one of the arcades under the Gloucester Building in Hong Kong which not only sold abacuses but also small books of instructions printed in English, I decided to find out just how difficult it really is to learn what the book called "Bead Arithmetic."

Anyone who has visited the Orient, and particularly Hong Kong and Kowloon, scarcely can fail to have been fascinated with the lightning speed and apparent accuracy displayed by any ordinary store clerk when he operates one of the abacuses which are to be found spaced along every store counter. It was noticed that in banks and other large business places where modern calculating machines were available these remained idle, for the most part, while the abacus beads clattered merrily. On inquiring why this was so I was told, "The abacus is much faster and is *never* wrong."

When I started to study my text book and attempted to learn how to perform the more simple operations of arithmetic on my abacus I found out several things. One was that the only text in English I could find in the bookstores was the one I had; that this book was sometimes confusing. Another was that a non-Oriental, no longer young in years, *could* learn to run an abacus—only slowly. Although the book, according to the Introduction, was written during the Yuan Dynasty (1280–1368 A.D.), without it I would have missed a considerable amount of pleasure and mental stimulation. When I got through with it I decided to write a small book of my own containing directions based, perhaps, more on non-Oriental philosophy of arithmetic and less on the Oriental insistence on memorizing and applying an interminable list of rules.

A considerable amount of pleasure may be obtained while learning to operate the abacus which is actually a very useful calculating machine and which may well have a place on many a non-Oriental desk.

March, 1958 W.J.C.

Contents

Introduction

The word "abacus" appears to have been derived from the Greek word "abax" which means a board, tablet, or calculating table. Some believe that the present day abacus, as used by the Chinese, known as *suan-p'an,* may be traced back to one of the earliest calculating devices—a board covered with dust.

In its present form, a frame containing columns of beads and having an appearance similar to that illustrated in the Frontispiece, with two beads above the crossbar and five below, the abacus appears to have been in use in China since the 13th century. The Japanese are said to have obtained a form of abacus from China, through Korea, in about the 16th century. The present day Japanese abacus, which they call *soroban,* a name derived from the Chinese *suan-p'an,* differs from the Chinese variety in several details. It is much smaller and sometimes has more columns, as many as thirty-one, but those with fewer columns are most often used. The essential difference between the Chinese and Japanese forms of the abacus is that the latter *has* only one bead per column above the crossbar and only four below, though some with one upper bead and five lower ones in each column can still be seen occasionally.

Some Fundamentals

(a) The Chinese abacus has two sets of beads, two above and five below, on each column. These sets are separated by a horizontal cross piece which extends across the frame. When the abacus is set ready to make a calculation all the beads above the bar, which we will call UPPER beads, are placed UP against the upper edge of the frame and all the beads below the bar, which we will call LOWER beads, are placed down against the lower edge of the frame.

(b) The abacus illustrated in this work has 13 columns, enough to perform most operations, but this number is not standard; abacuses with fewer or more columns may be obtained.

(c) Each UPPER bead, when moved down to the bar, represents five of its order and each LOWER bead when moved up, represents one of its order.

(d) On the first column from the right each LOWER bead equals one and each UPPER bead equals five. Column one is then a "unit" column. On column two, from the right, each lower bead equals ten and each upper bead equals fifty. Column two is, then, a "tens" column and so on. Figure 7 shows 6769—one upper and one lower bead ($5 + 1 = 6$) in the "thousands" column, one upper and two lower ($5 + 2 = 7$) in the "hundreds" column, one upper and one lower in the "tens" column and one upper and four lower ($5 + 4 = 9$) in the units column.

(e) Situations may arise during the operation of the abacus in which both upper beads on a column are already in the lowered position or all five lower beads are already in the raised position. If such situations are allowed to persist further operations of the abacus become impossible. The operation to remove such conditions will be referred to as "clearing".

As shown in *Figure 1,* column one has one upper bead in the lowered position and five lower beads in the raised position. Together they represent "10". Column two (tens column) has two upper beads in the lowered position and four lower beads raised; this column represents "140". The total number on the abacus is "150". However, both columns need clearing. The following steps for clearing this situation may be followed:

Fig. 1. Too many beads in action on a column. Need for "clearing."

(a) Column one: cancel (or lower) the five lower beads and compensate for this by lowering the second upper bead. The column still represents "10".

(b) Column one: cancel (or raise) the two upper beads and compensate by raising a lower bead on column two. In doing this, "10" is cancelled from column one and transferred to column two because one lower bead on column two equals "10".

(c) Column two now has all its upper beads lowered and all its lower beads raised. It is necessary to cancel (or raise) the two upper beads on column two ($50 + 50 = 100$) and compensate by raising one lower bead, on column three, which represents "100".

(d) We still have "150" represented on the abacus but all of the five lower beads on column two are raised and should be cleared. This is done by lowering all five beads and lowering one upper bead on the same (second) column.

We now have:	Column three, one lower bead	=	100
	Column two, one upper bead	=	50
	Column one, no beads in action	=	0
	Total		150

Figure 2 shows the abacus in the "cleared" condition, representing "150."

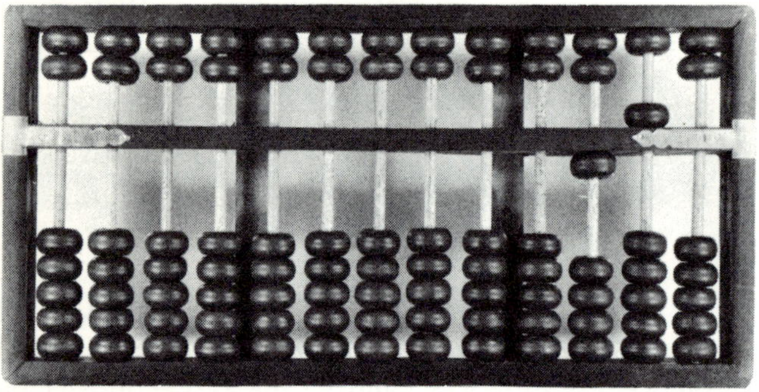

Fig. 2. Abacus in the "cleared" condition. Compare with Fig. 1.

As a general rule, whenever "clearing" is required, steps similar to the above should be carried out.

Bead Manipulation

The standard method of moving the beads on the abacus is to manipulate the *lower* beads with the thumb and index finger and the *upper* beads with the middle finger.

Calculations on the Abacus

There are four principal arithmetical calculations which may be performed on the abacus—addition, subtraction, multiplication, and division. Addition is the most common. The abacus can be used also for more complicated operations such as extracting the square root or the cube root of a number. All of these are illustrated in this manual.

In performing *addition* on the abacus a start should be made with figures on the *left,* whereas in written arithmetic the procedure is reversed. *Subtraction,* as in addition, is started from the left on the abacus. In *multiplication* the operation begins at the *right* and in *division* at the *left,* both similar to the practices in written arithmetic.

There is one most important and fundamental difference to be learned in the operation of the abacus and that is the method of thinking. When pencil arithmetic is performed one normally thinks of the total. For instance $9 + 8 = 17$. The abacus operator would have 9 on the board and in adding 8 would think of it as 10–2 and raise one LOWER bead on the "tens" column and lower two of the LOWER beads on the "units" (right hand) column. The answer on the abacus would then show 17, as illustrated in *Figure 3.* If the abacus operator wished to add 9 more to the 17 he would think of that step merely as adding 10 minus 1, by raising one lower bead on the "tens" column and lowering one lower bead in the "units" column. He would not think of it as $17 + 9 = 26$. In other words, his mental calculations are limited to a maximum of ten and each step is performed independently without being concerned with the total. The same is true with subtraction. In subtracting 8 from 17 he would think of taking away 8 from 10, leaving 2, and so he would lower one lower bead in the "tens"

Fig. 3. Result of adding 9 to 8 to make 17.

column and "raise" two lower beads in the "units" column, arriving at an answer of 9 on the abacus. No matter how many digits may be contained in the numbers to be added or subtracted, the entire operation is performed by applying this mechanical method to each digit in turn. Any attempt to calculate the answer mentally will retard the operation.

In problems of addition, if there are sufficient beads in a column to perform the operation, no problem is involved. For instance, in adding 7 to 12, the operator, after placing 12 on the board, merely moves down an upper (or 5 bead) and moves up two "unit" beads. The 5 and 2 make 7 and he gets the answer 19. Now if he desires to add 7 to 19, he does not think of 19 plus 7 totals 26 but rather he thinks merely of adding beads equivalent to 7, which in this case, inasmuch as there are not sufficient beads left in the "unit" column to add to 7, he automatically thinks of 7 being 10–3. He therefore raises one lower bead on the "tens" column and lowers three lower beads on the "unit" column. Again, his mental calculations are limited to a maximum of 10 *minus* a lesser number in the case of addition or cancelling

10 and *adding* a lesser number in the case of subtraction. Of course thêse operations become entirely automatic to a skilled abacus operator and he can execute them with great speed. In fact, abacus operators work so fast that when they compete with operators of our calculating machines, they often get the answer first.

The Chinese method of learning to operate the abacus depends on the memorizing of long lists of rules or "secrets" as the Chinese call them. Considering the great speed with which the abacus is operated in the Orient it is obvious that these lists of rules must be memorized so well that they come to mind subconsciously and just as readily as the fact that six times eight equals forty-eight comes to the non-Oriental mind. There are seventeen rules each for addition and subtraction and nine rules for division. This makes a total of 43 rules. It is suggested that these "rules" are very useful for *reference* when the student is first learning to operate the abacus, but after the rules and their application are understood then rote memorization will be unnecessary.

Addition

Following are the seventeen rules for addition. The Chinese say that to become a skillful abacus operator one must have a good memory and, above all, must practice everlastingly. In regard to the rules, they say that "these make the process of bead arithmetic easy and quick and therefore should be carefully and well committed to memory." The rules for addition apply where there are not sufficient beads in the lower part of the column to complete the operation. Rules (*a*) through (*d*) have to do with lowering a "5" bead and cancelling one or more unit beads; (*e*) through (*m*) are to be applied where a "10" unit must be raised and the difference cancelled from the "unit" column, and (*n*) through (*q*) where a 10 must be raised, a "5" bead in the "unit" column cancelled and beads in the "unit" column raised. It all looks very complicated at first but these rules are useful in showing the operations involved. As stated above, when their application is understood then rote memorization will be unnecessary.

Guides for Addition

a) One; lower five, cancel four
b) Two; lower five, cancel three
c) Three; lower five, cancel two
d) Four; lower five, cancel one
e) One; cancel nine, forward ten
f) Two; cancel eight, forward ten
g) Three; cancel seven, forward ten

h) Four; cancel six, forward ten
i) Five; cancel five, forward ten
j) Six; cancel four, forward ten
k) Seven; cancel three, forward ten
l) Eight; cancel two, forward ten
m) Nine; cancel one, forward ten
n) Six; raise one, cancel five, forward ten
o) Seven; raise two, cancel five, forward ten
p) Eight; raise three, cancel five, forward ten
q) Nine; raise four, cancel five, forward ten

The term "lower five" means to move down one upper bead. To "cancel" means to lower a *lower* bead from the middle beam and to "raise" means to move up a *lower* bead. To "forward ten" means to move up one lower bead on the next column to the left. In each of the rules, the first numeral indicates the number which is to be added to another number which is already on the frame; the remainder of the sentence explains how to carry this out.

Example I. Add 3 to 4

Place 3 on the frame by moving up three lower beads on the first column from the right. To this 4 is to be added. But three of the five lower beads on the column are already in use so we must *lower* an upper "5" bead and cancel one lower bead (see operation on Rule *d* above). The result is: one upper (or "5") bead and two lower beads on the abacus, representing "7", which is the answer.

Example II. Add 24 to 36 = 60

Place "24" on the abacus by raising four lower beads on the first column from the right and two lower beads on the second column from the right. To this is to be added 36.

It should be noted here that although, in written arithmetic we begin to add figures from the right, on the abacus we must do so from the left.

Therefore, we begin with the *second* column by lowering 5 and cancelling 2 (Rule *c*). On the *first* column we cancel 4 and forward 10 (Rule *j*). The answer is $5 + 1 = 6$, on the second column, which equals 60.

Example III. Add 783, 275, 14, 5697 and 34 = 6803

First place 783 on the frame *(Figure 4)*. Then starting with the *third* column from the right, add two by raising two of the lower beads. It will be noticed that neither Rule

Fig. 4. Example III in addition. Shows 783 set up on the abacus.

b nor Rule *f* can be applied. Now add 7 on column two by cancelling 3 and forwarding 10 (Rule *k*). Notice that all five of the lower beads on column three are raised. We should, therefore, cancel them all by lowering the second upper bead. Then cancel both those upper beads and raise one lower bead on the *fourth* column.

Now add 5 on the first column by lowering one upper bead. The abacus should now show: one lower bead raised on the fourth column, no beads on the third, one upper bead lowered on the second and one upper bead lowered plus three lower beads raised on the first column, representing 1058.

Fig. 5. Example III in addition. Shows the condition on the abacus after 275 has been added to 783, before clearing. Total 1058.

Figure 4 shows the number 783 set on the abacus. *Figure 5* illustrates the condition when 2 is added to the third column, 7 is added to the second column and 5 added to the first column. *Figure 6* shows the appearance of the abacus after "clearing". The number represented is $783 + 275 = 1058$.

Now proceed to add 14 by raising one lower bead on the second column. In order to add 4 to the first column we cancel 6 and forward 10 (Rule *h*). Now the abacus should show: column four, one lower bead raised; column three, no beads in operation; column two, one upper bead lowered and two lower beads raised; column one, two lower beads raised, representing 1072.

Fig. 6. Example III in addition. Shows the condition on the abacus after "clearing" the situation shown in *Fig. 5.*

Next add 5697 by first lowering one upper bead on the fourth column. Then lowering one upper bead and raising one lower bead on the third column (no rules apply). Now raise another lower bead on the third column and cancel one lower bead on the second column. This operatin is "cancel one forward ten" of Rule *m*. Finally lower one upper bead and raise two lower beads on the first

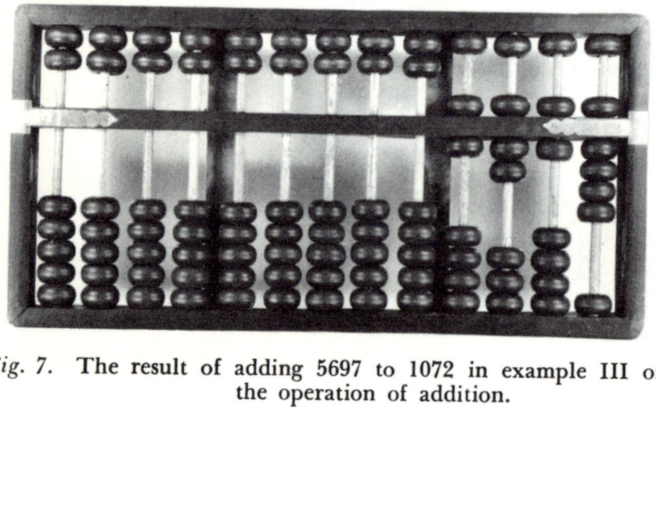

Fig. 7. The result of adding 5697 to 1072 in example III of the operation of addition.

column (no rule). The abacus should now look like *Figure 7*. There still remains to add 34. This may be accomplished by raising three lower beads on column two and lowering an upper bead and a lower one on the first column. The appearance of the abacus should be as shown in *Figure 8*. Here we have both "5" beads lowered on the first column.

Fig. 8. The result of adding 34 to 6769, before clearing. Total 6803.

Fig. 9. Final condition on the abacus at end of example III in addition. 6803.

This should be cleared by raising the two "5" beads on the first column and forwarding 10 by raising one lower bead on the second column. Now we have all five lower beads raised on the second column. This may be "cleared" by lowering these five beads and lowering one more upper bead on the second column. This leaves both upper beads lowered on the second column. This may be cleared by raising both these upper beads and also raising one lower bead on the third column. The answer is 6803 and the final condition on the abacus should appear as in *Figure 9.*

Example IV.

Try an addition involving the decimal point. Add the following: \$312.75, \$95.75, \$21.03 and \$2304.69 making a total of \$2734.22.

Start by placing \$312.75 on the abacus. The decimal point will lie between the second and third column. Now add \$95.75 by adding 9 to the fourth column. This may be done by cancelling the one lower bead on column four from the right and forwarding 10 (Rule *m*). Then add 5 to the third column by lowering one upper bead (no rule). Column two is a "cents" column (in this example). We wish to add 70 cents or 7 to column two by raising two lower beads, cancelling one upper bead and forwarding 10. Now 5 remains to be added to the first column. We can do this by cancelling 5 (raise the one upper bead) and forward 10 (raise one lower bead on column two). This latter operation results in all five lower beads on the second column being in the raised position. To clear this, lower the five lower beads and lower one upper bead. The sum of \$312.75 and \$95.75 should be \$408.50. The result of the addition thus far on the abacus is shown in *Figure 10.*

Fig. 10. Result on the abacus of adding $312.75 to $95.75. Equals $408.50. Example IV in addition.

The next sum to add is $21.03. Add 2 to the fourth column by raising two of the lower beads. Add 1 to the third column by raising one lower bead. There is nothing to add to the second column. Finally, add 3 to the first column by raising three lower beads. The total should now be $429.53 which should appear on the abacus as shown in *Figure 11.* Finally we must add $2304.69. It will be

Fig. 11. Example IV in addition. Shows $429.53 on the abacus.

necessary to add 2 to the sixth column by raising two of the lower beads. The number 3 is now added to the fifth column. This may be done by lowering an upper bead and also lowering two lower beads (Rule *c*). There is nothing to be added to the fourth column. The third column requires the addition of 4. This may be accomplished by applying Rule *h*, that is, cancelling one lower bead and one upper bead and raising one lower bead on column four. This last operation is "forward ten." Now we come to the "cents" columns which are the second and first. We first add 60 cents (out of the 69 required) to the second column by raising one lower bead and one upper bead (cancel five) and raising a lower bead on column three (forward ten) (Rule *n*). Then we add the remaining 9 cents to the first column. This is done easily by applying Rule M which says "Cancel one, forward ten." So we cancel one lower bead on the first column and raise one lower bead on column two. The total should be $2734.22 and this should appear on the abacus as shown in *Figure 12.*

Fig. 12. Example IV in addition. Total of $2734.22 on the abacus.

Example V.

Perform the following addition on the abacus:

	10363
	4596
	76969
	429
	7642
Total	99999

Place 10363 on the abacus and proceed to add 4596 which equals 14959, which should appear on the abacus as shown in *Figure 13.*

To the 14959 on the abacus we must add 76969. This means that the figure 7 must be added to the fifth column and so on. When this addition is made, the sum should appear on the abacus as shown in *Figure 14,* this is 91928.

The next addition is 429, which is started by adding 4 to the third column. $91928 + 429 = 92357$. This should appear on the abacus as shown in *Figure 15.*

Finally we must add 7642. When this addition is made the abacus should show 99999 as illustrated in *Figure 16.*

Fig. 13. Example V in addition. Result of adding 10363 and 4596. Equals 14959.

Fig. 14. Example V in addition. 91,928 as it should appear on the abacus.

Fig. 15. Example V in addition. 92,357 as it should appear on the abacus.

Fig. 16. Example V in addition. 99,999 as it should appear in the final answer.

Exercises

(a) Add 2.684 to 32.468. Answer 35.152
(b) Add 5.68435 to 37.85. Answer 43.53435
(c) Add 2.591 to 0.03902. Answer 2.63002
(d) Add 231.4 to 5.062. Answer 236.462
(e) Add 123,456,789 nine times which, if computed correctly, will give a total of 1,111,111,101.
(f) In order to attain some speed in abacus operation, have someone call out the following lists of figures:

(i)	(ii)	(iii)	(iv)
632	822	143	2480
468	821	268	7608
463	432	472	2581
210	268	432	9263
Total 1773	472	684	Total 21932
	Total 2815	722	
		Total 2721	

	(v)		(vi)		(vii)
	4572		411.9		25912
	91		43.8		873
	258		345.7		59716
	1926		29.1		77
Total	6847	Total	830.5		6097
					347
					2834
				Total	95856

Subtraction

Guides for Subtraction

As an aid to the solution of problems in subtraction we are faced again with 17 rules. These rules are the opposites of those given for addition. The Chinese advise "Memorize these guides well as that is the secret of an efficient abacist."

a) One; cancel five, return four
b) Two; cancel five, return three
c) Three; cancel five, return two
d) Four; cancel five, return one
e) One; cancel ten, return nine
f) Two; cancel ten, return eight
g) Three; cancel ten, return seven
h) Four; cancel ten, return six
i) Five; cancel ten, return five
j) Six; cancel ten, return four
k) Seven; cancel ten, return three
l) Eight; cancel ten, return two
m) Nine; cancel ten, return one
n) Six; cancel ten, return five, cancel one
o) Seven; cancel ten, return five, cancel two
p) Eight; cancel ten, return five, cancel three
q) Nine; cancel ten, return five, cancel four

The meanings of the words "cancel" and "return" may be made clear by solving a problem.

Example I.

Subtract 498 from 884. Answer 386. Set up 884 on the abacus so that 4 is shown on the first column from the right, 8 on the second column and 8 on the third column. Subtraction on the abacus, as in addition, is started from the left, that is, in this case, beginning with the 8 on the third column. The first operation will be to subtract 400 from 800. This may be accomplished by applying Rule *d,* to column three: "Cancel five, return one." In other words, cancel the upper bead and raise one lower bead. Column three now should have four lower beads raised, representing 400.

We now come to the second column from which 90 must be subtracted. This may be accomplished, conveniently, by applying Rule *m* which states: "Cancel ten, return one." To "cancel ten" one of the four lower beads in action on column three must be lowered and to "return one" one of the lower beads of column *two* must be raised. In this way

Fig. 17. Example I in subtraction. Shows result of subtracting 490 from 884. The reading is 394.

we have subtracted 490 from 884 and the abacus should read 394 as shown in *Figure 17.*

Finally we must subtract 8 from the 4 on the first column. The most convenient way to do this will be to apply Rule *p,* which states: "Cancel 10, return five, cancel three." To "cancel ten", lower one of the lower beads on column *two.* To "return five" lower one upper bead on column *one.* To "cancel three", lower three lower beads on column *one.* In other words −10 +5 —3 = −8. The final appearance of the abacus should be the same as shown in *Figure 18.*

Fig. 18. Example I in subtraction. Final appearance of the abacus. Answer 386.

Example II.

Find the difference between 3,402,731 and 1,924,005.

Set up 3,402,731 on the abacus and proceed to subtract 1 from 3 on the seventh column. This leaves two lower beads in the raised position on the seventh column. Now subtract 9 from 4 on the sixth column. It is possible to use Rule *m,* which states: "Cancel ten, return one". To cancel ten, lower one lower bead on column *seven;* to return one, raise one lower bead on the sixth column. The result is that all five lower beads on column six are raised. To clear them lower them and also lower one upper bead on the same column. It is possible also to apply Rule *q* which states: "Cancel ten, return five, cancel four" or $-10 + 5 - 4 = -9$. To cancel ten, lower one of the lower beads on column seven; to return five, lower one upper bead on column six; to cancel four, lower the four lower beads on column six.

So far we have subtracted 1,900,000 from 3,402,731 and the abacus should read 1,502,731, no matter whether we apply Rule *m* or Rule *q* and the frame should appear as shown in *Figure 19.*

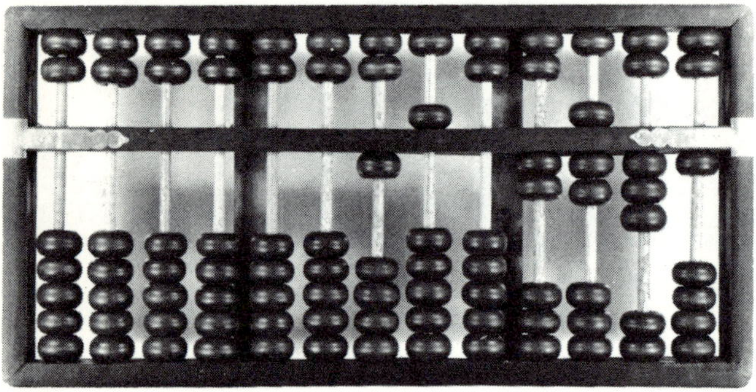

Fig. 19. Example II in subtraction. Shows result of subtracting 1,900,000 from 3,402,731. Result is 1,502,731.

The subtraction may be continued now so that, in the fifth column, 2 is subtracted from 0. Rule *f* may be applied. This states: "Cancel ten, return eight." To cancel ten, raise the upper bead on column *six* and raise four lower beads on the same column, $(-5+4 = -1)$; to return eight, lower one upper bead and raise three lower beads on column five (5+3 = 8). The abacus now should read 1,482,731 as shown in *Figure 20.*

We now move to the fourth column where 4 must be subtracted from 2. Rule *h* which states: "Cancel ten, return six" may be applied. To cancel ten, lower one lower bead

Fig. 20. Example II in subtraction. Abacus reading 1,482,731.

Fig. 21. Example II in subtraction. Abacus reading 1,478,731.

on column five; to return six, lower one upper bead and raise one lower bead on column four. The reading on the abacus now should be 1,478,731 as shown in *Figure 21*.

Since zeros are to be subtracted from columns three and two, no operation is involved. Therefore the residue remains 1,478,731. This brings us to column one where 5 must be subtracted from 1. Only one rule is applicable, namely Rule i, which states: "Cancel ten, return five." To cancel ten, lower one lower bead on column two; to return five, lower one upper bead on column one. Now the whole subtraction is completed and the abacus should read 1,478,726 as is shown in *Figure 22*.

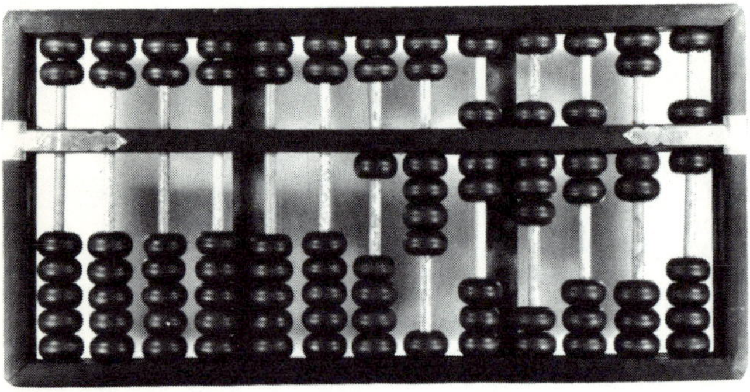

Fig. 22. Example II in subtraction. Final situation. Abacus reading 1,478,726.

Example III.

Here is an exercise having an Oriental flavor. It is taken from *The Fundamental Operations in Bead Arithmetic,* by C. C. Woo.

> If you buy at a Chinese grocery store tea for 78 cents, rice for 26 cents and sugar for as much difference

as the amounts paid for tea and rice, giving a two dollar bill, how much change should you receive?

This problem is solved, of course, by subtracting 26 cents from 78 cents, then adding that amount to 78 cents and 26 cents, or a total of $1.56, then subtracting that amount from $2.00.

So, to illustrate the last step, we raise two lower beads on the third column from the right to represent $2.00 and then proceed to subtract $1.56 from it. First we lower one lower bead on the third column. This leaves only one lower bead in the raised position. We now subtract 5 from 0 on the second column. We can do this by lowering the one lower bead on the third column and lowering an upper bead on the second column ($-100 + 50 = -50$). We have now 6 to subtract from 0 on the first column. This may be accomplished by cancelling ten and adding 4 ($-10 + 4 = -6$). Raise the upper bead on column two and raise four lower beads on the same column. Now raise four lower beads on column one. The answer is $0.44. (Four lower beads in raised position on both column one and two.)

It is suggested that the following example be worked out on the abacus, first by reference to the subtraction rules and then, later, without the rules and applying the ordinary principles of arithmetic. It is believed that the "rules" as such, should not be memorized, as advocated by some, but should be utilized by the beginner only for the purpose of learning how the abacus operates.

Example IV.

Subtract 34, 5697, 14, 275, and 783 from 6803.

First place 6803 on the abacus. Now subtract 3 from 0

on the second column. To do this, lower one lower bead on the third column and lower one upper bead and raise two lower beads on the second column (−10 + 7 = −3). Now subtract 4 from 3 on the first column. To do this lower one lower bead on column two and then lower one upper bead and raise one lower bead on column one. The beads on the abacus should represent 6769.

Now subtract 5697. Start with the fourth column. Subtract 5 from 6 by raising the upper bead. Then subtract 6 from 7 on column three by raising one upper head and lowering one lower bead. Next subtract 9 from 6 on column two. This is done by lowering one lower bead on column *three* and raising one lower bead on column two, (−10 + 1 = −9). Finally subtract 7 from 9 on column one by raising the upper bead and lowering two lower ones. The abacus should represent 1072.

Now subtract 14 from 1072. This may be done by subtracting 1 from 7 on column two, then subtract 4 from 2 on the first column. To accomplish this, make use of the equation −10 + 6 = −4 by lowering one lower bead on column two and lowering one upper bead and raising one lower bead on column one. The abacus should now represent 1058.

Now subtract 275 from 1058. Start with the third column by subtracting 2 from 0. This may be done by lowering the lower bead on the fourth column and lowering one upper bead and raising three lower beads on the second column (−10 + 8 = −2). Then move to the second column and subtract 7 from 5 by lowering one lower bead on column three and raising three lower beads on the second column. Finally, on the first column, subtract 5 from 8 by raising one upper bead. The abacus should now represent 783. When 783 is subtracted from this the answer should be zero, proving that our subtractions have been made correctly.

Example V.

The solution of the following example will require the application of the methods of both addition and subtraction.

$$
\begin{array}{r}
2530 \\
108 \\
-1703 \\
-\ \ 216 \\
2325 \\
-1072 \\
\hline
1972
\end{array}
$$

Place 2530 on the abacus and proceed to add 108. Starting with the third column add 1 by raising one lower bead. No addition is made to column two. Add 8 to column one by lowering one upper bead and raising three lower beads. The abacus should indicate 2638. The next number is 1703 and it is subtractive. Start at column four and subtract 1 by lowering one lower bead. Now move to the third column and proceed to subtract 7 from 6 by lowering the lower bead on column four and raising three lower beads on column three. There is no operation on column two. Finally, 3 must be subtracted from the 8 on column one by lowering three lower beads. The abacus should indicate 935. From this number 216 is to be subtracted. Start at the third column and subtract 2 from 9 by lowering two lower beads. Then subtract 1 from 3 on column two by lowering one lower bead. Finally subtract 6 from 5 on column one by lowering one lower bead on column two and raising four lower beads on column one. The abacus should now indicate 719. The next number is 2325 and is additive. This addition is started by raising two lower beads on the fourth column and then adding 3 to the third column by raising one lower bead on the fourth column and then on the third column raising one upper bead and lowering two lower beads (10−5−2 = 3). Moving to the second column, 2 must be added to 1 by raising two lower beads. It remains to add 5 to 9 on the first column. This may be done by raising one lower bead on column two and raising one upper bead on column one (10−5 = 5). The abacus should indicate 3044. The last number is 1072 and it is subtractive. Start at the fourth column. Subtract 1 from 3 by lowering one lower bead. There is no operation on column three. On column two, subtract 7 from 4. Here a more unusual situation arises. The abacus shows 2 lower beads raised on column four, no beads in action on column three, four lower beads raised on both columns two and one.

If we refer to the rules for subtraction it will be found that when 7 is to be subtracted the first operation is "cancel 10". However, in the present problem the third column has no beads in operation so that none can be cancelled. We can remedy this situation by adding 10 to column three by lowering one upper bead and raising five lower beads and then cancelling ten on column four by lowering one lower bead. Now we can proceed to subtract 7 from 4 on the second column by lowering one lower bead on column three, lowering one upper bead on column two and lowering two lower beads on column two (−10 + 5 −2 = −7). The last operation is to subtract 2 from 4 on column one by lowering two lower beads. The abacus should indicate 1972 which is the answer.

Exercises

(a) Subtract 2991 from 4963. Answer 1972.

(b) Subtract \$2.59 from \$64.00. Answer \$61.41.

(c) Subtract 0.00198 from 1.0675. Answer 1.06552.

(d) Find the sum of the following:

$$\begin{array}{r} 2.9063324 \\ -2.5628874 \\ 0.0180761 \\ -0.3605212 \end{array}$$

The answer is 0.0009999.

Multiplication

No table of "rules" is necessary as an aid in the performance of multiplication on the abacus. However, a knowledge of the multiplication tables, at least as far as 9 times 9, is mandatory.

The multiplicand (number to be multiplied) is placed on the right hand side of the abacus but the first column on the right cannot be taken as the unit column of the multiplicand. When the multiplier contains one figure, then the first column on the right must be left vacant. If the multiplier contains two figures, then the first two columns on the right must be left vacant, and so on. The number of vacant columns must equal the number of figures in the multiplier.

Suppose we wish to multiply 378 by 273. Since the multiplier (273) contains three figures the first three columns on the right hand side of the abacus must remain vacant. When the multiplicand (378) is placed on the abacus, the condition will be as shown in *Figure 23*. (The multiplier, 273, is shown on the left of the abacus.)

If there is room on the board the multiplier should be set up on the extreme left. This will put the complete problem on the board and avoid any chance of forgetting any of the numbers involved during its completion.

In multiplication, the operations begin from the *right,* though they begin from the left in addition and subtraction.

There is another fundamental difference in procedure between multiplication in abacus arithmetic and "pencil"

Fig. 23. Multiplication. 378 × 273. Placing multiplicand on the abacus. (The multiplier, 273, is shown on the left of the abacus.)

arithmetic. In "pencil" arithmetic when we multiply 378 by 3 (using Example I below) we do just that — 3 × 8, then 3 × 7, then 3 × 3. In abacus arithmetic after setting up 378 on the board the procedure is reversed, and 3 is multiplied by 8, then by 7, then by 3, as shown below.

Example I.

We will begin the study of multiplication with a simple case and then proceed to more complicated problems. The first problem: 3 × 378 = 1134, is to be solved as follows:

(a) Set up 378 on the abacus leaving the first column on the right vacant (one figure in the multiplier)

(b) 8 × 3 = 24. Place 4 on the first column, cancel 8 on the second column and replace it with 2. (It should be noted that the 8 being canceled was part of the problem as set up on the board originally, and is removed to make way for the answer to be put on the board. Its use in the problem as a multiplier is also completed. In the next two steps shown below, the 7 and the 3 will be eliminated likewise.)

(c) 7 × 3 = 21. (This is really 70 × 3 = 210) Add 1 to the second column, cancel 7 on the third column and replace it with 2

(d) 3 × 3 = 9 (3 × 300 = 900). Cancel 3 on the fourth column and add 9 to the third column (raise one lower bead on the fourth column and lower one lower bead on the third column). The abacus should represent 1134. In order that these operations may be followed closely, the various steps involved are tabulated as shown in *Figure 24*.

Step	Column	Beads		
		Upper	Lower	
(a)	4	—	3	= 300
Place 378 on	3	1	2	= 70
the Abacus	2	1	3	= 8
	1	—	—	0 378
(b)	4	—	3	
8 × 3 = 24	3	1	2	
	2	—	2	
	1	—	4	
(c)	4	—	3	
7 × 3 = 21	3	—	2	
	2	—	3	
	1	—	4	
(d)	4	—	1	= 1000
3 × 3 = 9	3	—	1	= 100
	2	—	3	= 30
	1	—	4	= 4 1134

Fig. 24. Example I in multiplication. Tabulation showing steps in multiplying 378 by 3.

Example II.

As a further step let us multiply 378 by 73. The product should be 27594. In doing this we will proceed to multiply 73 by 8, then by 7 and finally by 3. The following steps should be observed:

Fig. 25. Example II in multiplication. Condition on the abacus at the end of step (*a*).

Fig. 26. Example II in multiplication. Condition on the abacus at the end of step (*b*).

(a) Set up 378 on the abacus but see to it that the first two columns are left vacant because the multiplier contains two figures. (*Figure 25*)

(b) $8 \times 3 = 24$. Place 4 on the first column and 2 on the second column. (*Figure 26*)

Fig. 27. Example II in multiplication. Condition on the abacus at the end of step (*c*).

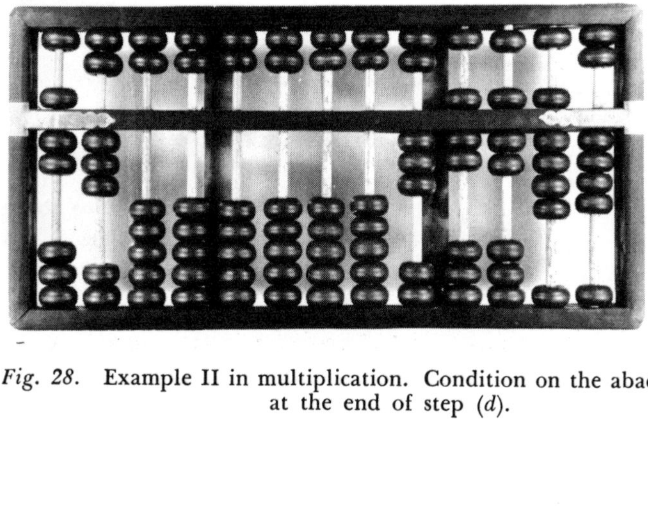

Fig. 28. Example II in multiplication. Condition on the abacus at the end of step (*d*).

(c) 8 × 7 = 56 (actually 8 × 70 = 560). Add 6 to the second column, and change 8 in the third column to 5. (*Figure 27*)

(d) 7 × 3 = 21 (really 210). Add 1 to the second column and 2 to the third column. (*Figure 28*)

(e) 7 × 7 = 49 (really 70 × 70 = 4900). Cancel 7 from the fourth column and replace it by 4. Add 9 to the third column. (*Figure 29*)

(f) 3 × 3 = 9 (really 900). Add 9 to the third column. (*Figure 30*)

(g) 3 × 7 = 21 (actually 70 × 300 = 21000). Cancel 3 in the fifth column and replace it with 2. Add 1 to the fourth column. The abacus should now indicate 27594 which is the answer. (*Figure 31*)

Fig. 29. Example II in multiplication. Condition on the abacus at the end of step (*e*).

Fig. 30. Example II in multiplication. Condition on the abacus at the end of step (*f*).

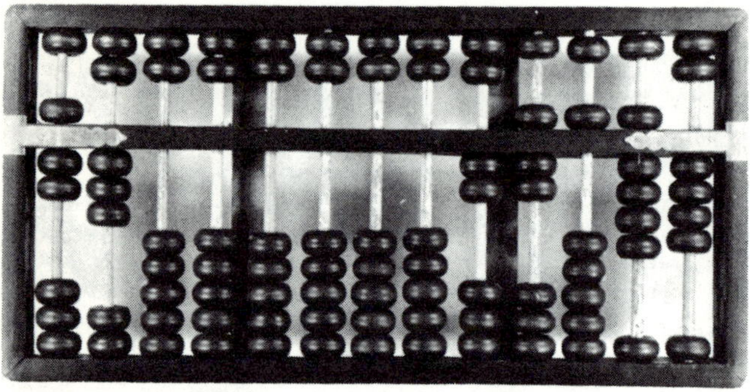

Fig. 31. Example II in multiplication. Condition on the abacus at the end of step (*g*).

Example III.

As a still further step, let us multiply 378 by 273 = 103194.

(a) Set up 378 on the abacus, leaving the first three columns on the right vacant.

(b) Multiply 3 by 378.

(i) 8 × 3 = 24. Place 4 on the first column and 2 on the second column. Cancel 8 on the fourth column.

(ii) 7 × 3 =21. Add 1 to the second column and two to the third column. Cancel 7 on the fifth column.

(iii) 3 × 3 = 9 (900). Add 9 to the third column. Cancel 3 on the sixth column. The abacus should show 1134.

(c) Multiply 7 by 378 (really 70 × 378) and add to the product of (b) (iii).

(i) 8 × 7 = 56 (70 × 8 = 560). Add 6 to the second column and 5 to the third column.

(ii) 7 × 7 = 49 (70 × 70 = 4900). Add 9 to the third column and 4 to the fourth column.

(iii) 3 × 7 = 21 (70 × 300 = 21000). Add 1 to the fourth column and 2 to the fifth column. The abacus should show 27594.

(d) Multiply 2 × 378 and add to the product of (c).

(i) 8 × 2 = 16 (actually 1600). Add 6 to the third column and 1 to the fourth column.

(ii) 7 × 2 = 14. Add 4 to the fourth column and 1 to the fifth column.

(iii) 3 × 2 = 6. Add 6 to the fifth column. The multiplication of 378 by 273 is now completed and the abacus should show 103194.

Further practice on multiplication on the abacus should be had by attempting the solution of the following:

287 by 41 = 11767
601 by 458 = 275,258
231.4 by 5.062 = 1171.3468
35.86 by 0.00294 = 0.1054284

Division

Division may be defined as the process of finding out how many times a number or quantity is contained in another. It is the reverse of multiplication. The number to be divided is called the "dividend" and the number by which the dividend is to be divided is called the "divisor." If there is a part of the dividend left over after the division operation is completed, it is called the "remainder." The number of times the divisor is contained in the dividend is called the "quotient."

"Short" division is the process used when the divisor has a value of less than 10. "Long" division is the process used when the divisor is greater than 10.

Guides for Short Division

The following guides or rules are given for use in short division. Like the "rules" for addition and subtraction, these rules for division should be understood rather than memorized.

a) *Divided by One.* $1 \div 1$, forward one, $2 \div 1$, forward 2, $3 \div 1$, forward 3, and so on. By "forwarding one" we mean to move up one lower bead on the left column next to the figure divided.

b) *Divided by Two.* $1 \div 2 = 0.5$, $2 \div 2$ forward 1. $4 \div 2$ forward 2 and so on. In the case of $1 \div 2 = 0.5$, simply change the dividend, that is 1, into 5 on the same column. Ordinarily, the unit place in the quotient is on the column to the left of the dividend figure. In other words, when 1 is divided by 2, the quotient is one-half or $\frac{5}{10}$ or 0.5.

c) *Divided by Three.* 1 ÷ 3 = 3 + 1. 2 ÷ 3 is 6 + 2. (20/3 = 6 + 2). 3 ÷ 3 = forward 1. 9 ÷ 3 = forward 3. The guides for dividing by three are explained in the same way as for dividing by two, except that the number after the plus, which is the "remainder", is to be *added* to the dividend figure, if any, on the column to the *right* of the quotient figure.

d) *Divided by Four.* 1 ÷ 4 = 2 + 2 (in other words: 10/4 = 2 with a remainder of 2). 2 ÷ 4 = 5. 3 ÷ 4 = 7 + 2 (30/4 = 7 + 2 remainder) and so on for five, six, seven, eight and nine. For instance, 8 ÷ 9 = 8 +8. All this appears to be confusing but the confusion can be cleared up readily by considering, in detail, a few examples of division operations on the abacus.

Example I.

11767 divided by 2 equals 5883.5.

(a) Place 11767 on the left side of the abacus. The operation of division on the abacus begins on the left and in this respect differs from multiplication which begins on the right. Therefore, in division, we will *number the columns from the left.* It is often advisable to keep the first column on the left side of the abacus empty for division operations, although it is not necessary to do so for the examples given here. After the number 11767 has been placed on the abacus, the situation will appear as shown in *Figure 32.*

Fig. 32. Example I in division. Setting up the dividend on the abacus when 11,767 is divided by 2.

(b) Starting with the first column to the left, convert the 1 into 5 ($1 \div 2 = 5$).

(c) Convert the 1 in the second column into 5.

(d) Cancel out 6 from the 7 on column three leaving 1 and advance 3 to column two; then convert the 1 remaining on column three to 5. Or, this step might be considered in another way. The problem is $7 \div 2 = 3 + 1$. We might consider this to be $(6 \div 2) + (1 \div 2)$. Cancel out 6 from column three leaving 1. Advance 3 to column two. We now deal with the (residual) 1 on column three. $1 \div 2 = 5$. Replace 1 on the column three with 5. The first procedure given is orthodox.

(e) On column four, $6 \div 2 = 3$. Cancel the 6 and advance 3 to column 3.

(f) On column five, $7 \div 2 = 3 + 1$. Cancel out 7 and replace it with 1, advance 3 to column four; now convert the "1" remaining on column five to "5".

The abacus should show 58835 which should be interpreted as 5883.5, the correct answer.

Example II.

This will be a much longer exercise. Divide 123,456,789 by 5. The answer should be 24,691,357.8.

(a) Place 123,456,789 on the left side of the abacus, with the figure "1" placed on the first column. Proceed to divide by 5.

(b) First column. 1 ÷ 5 = 2. Really 10/5 = 2. Cancel 1 on column one and replace it with 2.

(c) Second column. 2 ÷ 5 = 4 (20 ÷ 5 = 4). Replace 2 on column two with 4.

(d) Third column. 3 ÷ 5 = 6 (30 ÷ 5 = 6). Replace 3 on column three with 6.

(e) Fourth column. 4 ÷ 5 = 8. Replace 4 on column four with 8.

(f) Fifth column. 5 ÷ 5 = 1. (Here we have a quotient which is a whole number (1) and this will be "advanced" to column four. There is no residue.) Cancel 5. Forward 1.

(g) Sixth column. 6 ÷ 5 = 1 + 1. Cancel 5 leaving 1 residue. Forward 1 to column five. We now have to dispose of the residual 1 on this column. 1 ÷ 5 = 2. Replace the residual 1 with 2.

(h) Seventh column. 7 ÷ 5 = 1 + 2. Cancel 5 and forward 1 to column six. This leaves a residual of 2 on this column. 2 ÷ 5 = 4. Change the residual 2 to 4.

(i) Eighth column. 8 ÷ 5 = 1 + 3. Cancel 5 leaving a residual of 3. Advance 1 to column seven. Deal with the residual 3 ÷ 5 = 6. Change the residual 3 to 6.

(j) Ninth column. 9 ÷ 5 = 1 + 4. Cancel 5 leaving a residual of 4. Advance 1 to column eight. Dispose of the residual 4. 4 ÷ 5 = 8. Change 4 to 8 on column nine. The abacus should indicate 246,913,578 which should be interpreted as 24,691,357.8.

From the experience gained from the preceding examples in division it will have been noticed that when the quotient obtained is a whole number, for instance 5 ÷ 2 = 2 + 1, then this quotient number (2) is "advanced" which means it is added to the preceding column. However, when the quotient is obtained by multiplying the dividend by 10, for instance, 4 ÷ 8 = 5 (40 ÷ 8 = 5) then the quotient 5 is placed in the column in which the dividend occurs, after cancelling out the dividend. When a residual remains (in 5 ÷ 2 = 2 + 1, 1 is the residual), the residual is added to the column next to the right of that in which the dividend occurred.

Example III.

Divide 123,456,789 by 9.

(a) Set up 123,456,789 on the left of the abacus, with the number 1 on column one.

(b) First column. 1 ÷ 9 = 1 + 1. Leave 1 on column one and add 1 to column two (column two now contains 3, the residual 1 having been added to the column to the right (2 + 1 = 3).

(c) Second column. 3 ÷ 9 = 3 + 3. Leave 3 on column two and add 3 to column three which now shows 6.

(d) Third column. 6 ÷ 9 = 6 + 6. Leave 6 on column three and add 6 to column four. (Column four now contains 10, having one upper bead lowered and five lower beads raised. Allow this condition to *remain* for the moment.)

(e) Fourth column. 10 ÷ 9 = 1 + 1. Advance 1 to column three and leave 1 residual on column four. Dispose of this residual. 1 ÷ 9 = 1 + 1. Leave 1 on column four and add one to column five.

(f) Fifth column. 6 ÷ 9 = 6 + 6. Allow 6 to remain on column five and add 6 to column six. Column six now contains 12, having two upper beads lowered and two lower beads raised. Allow this condition to remain for the moment.

(g) Sixth column. 12 ÷ 9 = 1 + 3. Cancel 12 from column six and replace it with 3. Forward one to column five. Dispose of the residual 3 on column six. 3 ÷ 9 = 3 + 3. Leave 3 on column six, add 3 to column seven. Column seven now contains 10, consisting of one upper bead lowered and five lower beads raised. Allow this situation to remain for the moment.

(h) Seventh column. 10 ÷ 9 = 1 + 1. Subtract 9 from column seven, leaving 1. Add 1 to column six. Dispose of the residue 1 on column seven. 1 ÷ 9 = 1 + 1. Leave 1 on column seven, add one to column eight which now contains 9.

(i) Eighth column. 9 ÷ 9 = 1. Cancel 9 on column eight. Advance 1 to column seven.

(j) Ninth column. 9 ÷ 9 = 1. Cancel 9 on column nine and advance 1 to column eight. The abacus should now indicate 13,717,421, which is the correct answer.

Examples I, II and III should be practiced many times with the aid of these notes and then without the notes.

Long Division

When the divisor contains more than one figure, the procedure on the abacus is changed considerably and it is somewhat more complicated than that for short division. The operations involved in long division may be summarized into three general procedure steps dealing with:

(a) Short division
(b) Multiplication
(c) Subtraction

Each of these steps now will be discussed. Having placed the dividend on the left of the abacus, proceed as follows:

(a) *Divide* the first numeral of the dividend by the first numeral of the divisor and obtain a partial quotient.
(b) *Multiply* the partial quotient, obtained from operation (a), by the rest of the divisor.
(c) *Subtract* the product of operation (b) from the dividend.

Example I.

Divide 112 by 56.

(a) Set up 112 on the left hand side of the abacus.
(b) *Divide* the first numeral of the dividend, which is 1, by the first numeral of the divisor, which is 5. $1 \div 5 = 2$ (really $10 \div 5 = 2$). Add 1 to column one, making 2 on that column. 2 is then the partial quotient.
c) *Multiply* the partial quotient, which is 2, by the remainder of the divisor, which is 6. $2 \times 6 = 12$.
(d) *Subtract* the product of operation (c), which is 12, from the residue of the dividend, which is 12. $12 - 12 = 0$. Cancel 1 from column two and cancel 2 from column three. The figure 2 remains on column one and this is the quotient and answer.

Rules for Long Division

Again a set of rules or guides is offered for use while the student is learning to operate long division on the abacus. When the logic of the method is understood, then the use of the guides should be discontinued. It may be true that Orientals, having memorized all these rules, can apply them subconsciously. If this is so, then one should begin to learn to operate the abacus in early childhood and then practice its use every day for the remainder of one's life, if one desires to attain high speed skill.

It may be pointed out that, according to the "rules", if the left hand figure of the dividend is equal to that of the divisor, it would leave no minuend from which the product of the quotient, multiplied by the remainder of the divisor, must be subtracted or when the minuend is too small for the product to be subtracted. The word "minuend" means the number from which another number is to be subtracted.

a) *Division by One.* 1 ÷ 1, if there is no minuend, is 9 + 1. That is, to change the dividend figure 1 into 9 and add 1 to the next column to the right. If the minuend is *insufficient,* subtract 1 from the quotient figure and add 1 to the dividend figure on the next column to the right. This operation is "subtract one, return one".

b) *Division by Two.* 2 ÷ 2, if no minuend, is 9 + 2. Insufficient, subtract 1, return 2.

c) *Division by Three.* 3 ÷ 3, if no minuend, is 9 + 3. Insufficient, subtract 1, return 3.

d) *Division by Four.* 4 ÷ 4, if no minuend, is 9 + 4. Insufficient, subtract 1, return 4.

e) *Division by Five.* 5 ÷ 5, if no minuend, is 9 + 5. Insufficient, subtract 1, return 5.

f) *Division by Six.* 6 ÷ 6, if no minuend, is 9 + 6. Insufficient, subtract 1, return 6.

g) *Division by Seven.* 7 ÷ 7, if no minuend, is 9 + 7. Insufficient, subtract 1, return 7.

h) *Division by Eight.* 8 ÷ 8, if no minuend, is 9 + 8. Insufficient, subtract 1, return 8.

i) *Division by Nine.* 9 ÷ 9, if no minuend, is 9 + 9. Insufficient, subtract 1, return 9.

Example II.

Divide 280 by 56. Answer 5. Set up 280 on the left hand side of the abacus.

(a) *Divide* the first numeral of the dividend, which is 2, by the first numeral of the divisor, which is 5. 2 ÷ 5 = 4 (20 ÷ 5 = 4). The partial quotient is 4. Add 2 to column one to make 4 on that column.

(b) *Multiply* the partial quotient, which is 4, by the remainder of the divisor, which is 6. 4 × 6 = 24.

(c) *Subtract* the product of (b), which is 24, from the remainder of the dividend, which is 80 (8 on column two and 0 on column three). 80 − 24 = 56. The residue of the dividend is 56. Cancel 3 from column two, leaving 5 and add 6 to column three. Repeat the process on the residue, which is 56.

(a–i) *Divide* the first numeral of (residue of) the dividend, which is 5, by the first numeral of the divisor, which is also 5. 5 ÷ 5 = 1. Add 1 to column one, making 5 on that column. Cancel 5 from column two. The partial quotient is 1.

(b–i) *Multiply* the partial quotient, which is 1, by the remainder of the divisor, which is 6. 1 × 6 = 6.

(c–i) *Subtract* the product from (b–i), which is 6, from the remainder of the dividend, which is 6 (on column three). 6 − 6 = 0. Cancel 6 from column three.

5 remains on the abacus and is the quotient and answer.

Example III.

Divide 504 by 56. Answer: 9.

Place 504 on the left hand side of the abacus.

(a) *Divide* the first numeral of the dividend, which is 5, by the first numeral of the divisor, which is also 5. $5 \div 5 = 9 + 5$ (see rules for long division). Add 4 to column one, making 9 on that column and add 5 to column two. The partial quotient is 9.

(b) *Multiply* the partial quotient, which is 9, by the remainder of the divisor, which is 6. $9 \times 6 = 54$.

(c) *Subtract* the product of (b), which is 54, from the remainder of the dividend, which is also 54. $54 - 54 = 0$. Cancel 5 on column two and 4 on column three. There remains on the abacus only 9 on column one and 9 is the quotient and the answer.

Example IV.

Divide 102816 by 136. Answer: 756.

Place 102816 on the left hand of the abacus leaving column one vacant. In this problem the divisor contains three numerals.

(a) *Divide* the first numeral of the dividend, which is 1, by the first numeral of the divisor, which is also 1. $1 \div 1 = 9 + 1$. Cancel 1 on column two and place 9 on column one. Add 1 to column three. The partial quotient is 9.

(b) *Multiply* the partial quotient, which is 9, by the *second* numeral of the divisor, which is 3. $9 \times 3 = 27$. But this number (27) is greater than the minuend which is 12. (First two numerals of the remainder of the dividend, represented by 1 on column three and 2 on column four.)

We must, therefore, subtract 1 from the partial quotient, which is 9, leaving 8 on column one and add 1 to column three, making 2 on that column. 8 is now the new (trial) partial quotient. When this new quotient of 8 is multiplied by 3 (second figure of the divisor), the product is 24; but again this figure of 24 is greater than the minuend which is 22 (2 on column three and 2 on column four). Again we must subtract 1 from column one and add 1 to column three. The new partial quotient is 7 and 7 multiplied by 3 equals 21. This number is smaller than the minuend, which is 32 (3 on column three and 2 on column four).

(c) *Subtract* the product of (b), which is 21, from the first two numerals of the residue of the dividend. The number is 32 and is represented by 3 on column three and 2 on column four. 32 − 21 = 11. Cancel 2 from column three, leaving 1 and cancel 1 from column four, leaving 1.

(b–i) *Multiply* the partial quotient, which is 7, by the third (and remaining) numeral of the divisor, which is 6. 7 × 6 = 42.

(c–i) *Subtract* the product of (b–i), which is 42, from 118 which represents the first three figures of the residue of the dividend (1 on column three, 1 on column four and 8 on column five). 118 − 42 = 76. Cancel 1 from column three, leaving 0. Add 6 to column four to make 7 and cancel 2 from column five, leaving 6. The undivided residue of the dividend is now 7616. The next step is to divide the residue of the dividend, which is 7616, by the divisor, which is 136.

(a) *Divide* the first numeral of the residue of the

dividend (see (c–i) above), which is 7, by the first numeral of the divisor, which is 1. $7 \div 1 = 7$. The (trial) partial quotient is 7. Place 7 on column two, cancel 7 from column four. When the (trial) quotient is multiplied by 3, which is the second figure of the divisor, the product is 21. This number is greater than the minuend which is 6. Trial will show that 5 must be used as the partial quotient. Cancel 2 from column two, leaving 5 (new partial quotient). Add 2 to column four, making 2 on that column.

(b) *Multiply* the partial quotient, which is 5, by the second figure of the divisor, which is 3. $5 \times 3 = 15$.

(c) *Subtract* the product of (b), which is 15, from the first two figures of the residue of the dividend, which is 26 ($26 - 15 = 11$). Cancel 1 from column four, leaving 1, and cancel 5 from column five, leaving 1.

(b–i) *Multiply* the third (and last) figure of the divisor, which is 6, by the partial quotient, which is 5. $6 \times 5 = 30$.

(c–i) *Subtract* the product of (b–i), which is 30, from the residue of the dividend, which is 111 (columns four, five and six). $111 - 30 = 81$. Cancel 1 on column four, leaving 0, add 7 to column five, making 8. The residue of the dividend is now 816 which remains to be divided. The condition on the abacus should appear as shown in *Figure 33*. Now there remains to be divided the residue 816 by the divisor, which is 136. The residue appears on columns five, six and seven.

(a) *Divide* the first numeral of the residue of the dividend, which is 8, by the first numeral of the

Fig. 33. Example IV in long division. Conditions on the abacus after some operations (c–i) in dividing 102,816 by 136.

divisor, which is 1. 8 ÷ 1 = 8. However, if 8 is used as a partial quotient and multiplied by 3, which is the second figure of the divisor, the product, 24, is greater than the minuend, which is 16. It will be found by trial that a partial quotient of 6 should be used. Cancel 6 from column five, leaving 2. Place 6 (partial quotient) on column three.

(b) *Multiply* the partial quotient, which is 6, by the second numeral of the divisor, which is 3. 6 × 3 = 18.

(c) *Subtract* the product of (b), which is 18, from the first two figures of the residue of the dividend, which is 21. 21 − 18 = 3. Cancel 2 from column five and add 2 to column six, making 3 on that column. The final residue is 36. The last step is to multiply the last partial quotient by the third and last numeral of the divisor and subtract this product from the residue of the dividend.

(b–i) *Multiply* the last partial quotient, which is 6,

by the final numeral of the divisor, which is 6. $6 \times 6 = 36$.

(c–i) *Subtract* the product of (b–i) which is 36, from the undivided residue of the dividend, which is 36. $36 - 36 = 0$. Cancel three from column six and cancel 6 from column seven. There remains on the abacus 756 which is the quotient and the answer. These figures should be in the first three columns.

Exercises:

(a) Divide 63,246,846 by 3. The answer is 21,082,282.
(b) Divide 242,051 by 53. Answer is 4567.
(c) The quotient is 756 and the dividend is 102,816. What is the divisor? Answer 136.

Square Root

The extraction of the square root of a number may be accomplished easily enough by means of the abacus but the process is somewhat slow and tedious. It consists of a long series of additions and subtractions. The operation can be performed, no doubt, much more expeditiously by making use of a table of logarithms, even when competing with a fast and skillful abacus operator.

To Extract the Square Root

The general steps in extracting the square root with the aid of the abacus may be stated as follows:

(a) Set up the number, the square root of which is to be extracted, on the *right* hand side of the abacus leaving a number of columns vacant, depending on whether or not the number is a perfect square and whether or not several decimal places are required in the answer. The number, the square root of which is to be extracted, is called the "square" number. Starting at the decimal point it should be "pointed off" to form groups containing two numerals, in a manner similar to that performed in pencil and paper arithmetic.

On the *left* hand side of the abacus, the number 1 is placed on column one. This number, on the left, is called the "root" number.

Beginning with the left hand group of the "square" number, subtract the number 1. Then add 2 to the 1 of the "root" number, making 3 and subtract 3 from the residue of the left hand group of the square number. Continue adding 2 to the root number and subtracting the sum from the residue of the left hand group until the root number becomes larger than the residue of the left hand group of the square number. When this occurs add a zero to the root number and add 11. Now include the second group of two numerals with the residue of the first group of the square number and subtract from them the augmented root number. Continue adding 2 to the root number and subtracting from the square number as before and until the original square number is exhausted. If the square is not a perfect square and when the decimal point has been reached, further groups of 00 may be added and the process continued.

To the root number, which finally has been obtained, add 1 and divide by 2. The quotient will be the square root sought. The process of extracting the square root is much easier demonstrated by means of examples.

Example I.

$\sqrt{484} = 22$

(a) Place 484 at the *right* side of the abacus, leaving column one vacant. Separate 484, the "square" number, into two groups, the first group containing one figure, 4, on column four and the other group containing two figures, 8 and 4 on column three and column two respectively.

(b) Place 1 on the first column on the *left* hand side of the abacus. This is the "root" number.

(c) Subtract 1 from the figure of the first "square" group, namely 4, (column four) leaving 3.

(d) Add 2 to the root number (1) making 3 on the (left hand) column one. Subtract 3 from (right hand) column four, leaving zero. The first square group is exhausted.

(e) Add 0 to the 3 in the "root" number and then add 11. 30 + 11 equals 41, which is the new root number. Left hand column one contains 4 and left hand column two contains 1. We now have 84 in the "square" number in columns three and two from the right. Subtract 41 from 84 leaving 43 in the "square" number.

(f) Add 2 to the 41 in the "root" number, making 43. Subtract "root" number 43 from "square" number 43 leaving zero.

(g) Add 1 to the "root" number, making 44. Divide 44 by 2 and obtain 22. This is the square root required.

Example II.

$\sqrt{974169} = 987.$

(a) Place 974169 on the right of the abacus and leave one column vacant. Mark off this "square" number in groups of two numerals. The first group will be 97, the second 41, and the third 69. Place 1 on column one on the left of the abacus. This is the "root" number.

(b) Proceed as in Example I. Subtract 1 from 97, leaving 96. Add 2 to 1 to make 3 in the "root" number. Subtract 3 from 96 to leave 93 in the "square" number and so on, adding 2 to the "root" number and subtracting from the "square" number. Finally the "root" number becomes 17 and the number in the first group of the square number is only 16. Now add 0 to the "root" number, making 170 and then add 11, making 181, which is the new "root" number. Attach the second group, 41, of the square number to the 16 remaining in the first group. The new "square" group is now 1641.

(c) Subtract 181 from 1641 leaving 1460. Add 2 to 181

making 183 for the "root" number. Subtract 183 from 1460 leaving 1277 and so on until the "root" number is 195 and the "square" number is only 137.

(d) Add 0 to the "root" number 195 to make 1950 and then add 11, making 1961. Attach the third group, 69, of the "square" number to the remaining 137, making 13769. Subtract 1961 from 13769 leaving 11808. Continue adding 2 to the "root" number and subtracting from the "square" number. Finally, when the "root" number has become 1973, the "square" number has become exhausted.

(e) Add 1 to the "root" number 1973 to make 1974 and divide by 2. 1974 ÷ 2 equals 987 which is the root required.

In connection with the operation of adding a zero to the "root" number and then adding 11, it may happen that when the next group of the "square" number is attached, the root number still may be too large. In such a case, add another group but instead of adding only one zero to the root number, add two zeros and then add 101 instead of 11.

Exercises:

(a) $\sqrt{30.714} = 5.542$

(b) $\sqrt{327.12} = 18.087$

Cube Root

The extraction of the cube root of a number on the abacus is more complicated but not more tedious than the extraction of the square root. However, it can be accomplished and the method is of interest, perhaps, from this point of view. As far as practicability is concerned, the use of logarithmic tables is much to be preferred.

The general procedure for extracting cube root is as follows:

(a) Place the "cube" number on the right side of the abacus and point it off, starting at the decimal point, into groups each containing *three* numerals.

(b) Place 1, on column one, on the left hand side of the abacus. This is the "root" number.

(c) Place 1 on column six (from the left) and call it the "square" number. We now have three sets of numbers, from left to right, the "root" number, the "square" number and the "cube" number.

(d) Subtract the 1 of the "square" number from the first or left hand group of the "cube" number.

(e) Add 1 to the "root" number and add the sum to the "square" number.

(f) Add an additional 2 to the "root" number and add the sum to the "square" number.

(g) Subtract the sum of the "square" number from the residue of the "cube" number and repeat the process given in (f) and (g) until the number in the first group of the

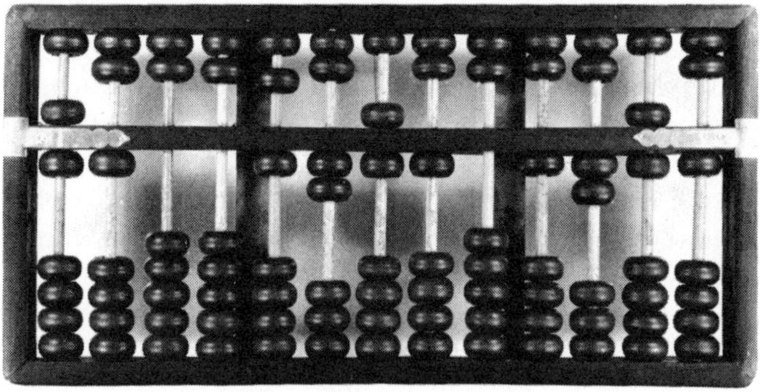

Fig. 34. Example I in cube root. Condition on the abacus after operation (*g*) when extracting the cube root of 9261.

"cube" number is less than the "square" number. *Figure 34.*

(h) Now attach the second group (three numerals) of the cube number, to the residue of the first group. Add 1 to the "root" number and add the sum of the "root" number to the square number. Then add 0 to the "root" number and add 11. Add the sum (total root number) to the "square" number but on the second column to the right, from the left hand figure of the "square" number. For example: If the total "root" number is 61 and the square number is 12, the new square number will be 1261.

(i) Subtract the new "square" number from the "cube" number and proceed as before until the "cube" number is exhausted.

(j) If, after attaching a new group of the "cube" number, the "square" number is still too large, attach an additional group but instead of adding 0 to the "root" number, followed by the addition of 11, in this case add 00 followed by the addition of 101. The sum is then added to the "square" number on the *fourth* column to the right from its left hand figure.

(k) When the "cube" number has been exhausted, add 2 to the final "root" number and divide by 3. The quotient is the cube root sought.

If the number, whose cube root is sought, consists of five or more numerals, an abacus with more than thirteen columns will be required. For instance, when extracting the cube root from a number containing five numerals, the columns occupied on the abacus will be:

By root number	3
By square number	4
By cube number	5
Total	12

With a 13-column abacus, there will not be enough vacant columns between the various sets of numbers. This would lead to confusion.

Example I.

$$\sqrt[3]{9261} = 21.$$

(a) Place 9261 on the right side of the abacus. This is the "cube" number. Place 1 on column one on the left hand side of the frame. This is the "root" number. Place 1 on a column near the center of the abacus, say, on column six. This is the "square" number.

(b) Point off the "cube" number, starting at the decimal, into groups containing three numerals. In the case of 9261, the first or left hand group will consist of the number 9. The second group will consist of the numbers 261.

(c) Subtract the "root" number, 1, (from first group of cube number) leaving 8.

(d) Add 1 to the "root" number (1 + 1 = 2) and add

this to the 1 of the "square" number (1 + 2 = 3). Change one on column six from the left to 3. The "root" number is now 2. The "square" number is now 3 and the "cube" number is 8 (on column four from the right).

(e) Add 2 to the "root" number, 2 + 2 = 4. Add the sum of the root number, 4, to the "square" number, 3 + 4 = 7. This makes 7 on column six from the left. Subtract the sum of the "square" number, namely 7, from the "cube" number, 8 − 7 = 1, leaving 1 on column four from the right. The "square" number is now larger than the residue from the first group of the "cube" number.

(f) Attach the second group of the "cube" number to the residue of the first group. The "cube" number is now 1261. Add 1 to the "root" number, 4 + 1 = 5. Add the sum of the "root" number, namely 5, to the "square" number, which is 7. 5 + 7 = 12.

(g) Add 0 to the "root" number, making 50 and add 11, making 61. This is the new "root" number. Attach the numerals 6 and 1 to the "square" number to form 1261. The situation on the abacus is shown in Figure 34, in which, numbering the columns from the left, the "root" number, 61, occupies columns one and two; the "square" number, 1261, occupies columns five, six, seven and eight; the "cube" number, 1261, occupies columns ten, eleven, twelve and thirteen.

(h) Subtract the "square" number from the "cube" number. 1261 − 1261 = 0. There remains on the abacus only the "root" number, namely 61. Add 2 to the "root" number and divide by 3. 61 + 2 = 63. 63 ÷ 3 = 21. 21 is the cube root sought.